This book belongs to:

Let's Plant!

Sandy Mahony & Mary Lou Brown

Rain is needed for growing flowers and plants!

Sun is needed for growing flowers and plants!

Soil, water, and sun are needed to grow flowers and plants!

What do you need in order to grow plants?

Sun, water, and soil!

Fruit

ACROSS

DOWN

Vegetables

Answers to Crossword Puzzles

FRUIT

VEGETABLES

ACROSS
1. Apple
2. Lemon
5. Orange
9. Apricot
10 Kiwi
11. Pomegranate
12. Coconut

DOWN
1. Avocado
3. Grapes
4. Persimmon
6. Papaya
7. Pineapple
8. Lychee

1. Pepper
2. Eggplant
3. Cauliflower
4. Tomato
5. Carrot
6. Broccoli
7. Leek
8. Peas
9. Radish

Adventure Learning Press

adventurelearningpress.com